SJ Teacher's Choice Math Series

Algebra Practice Sets:
100 Problems & Solutions
(Volume 1)

For Grade 9 & 10 Students

SANJAY JAMINDAR

PREFACE

The aim of "**Algebra Practice Sets: 100 Problems & Solutions (Volume 1)**" book is to help primary school students of Grade 9 and 10 (Class-IX, X) develop their Algebra problem solving skills and expand their knowledge of basic Algebra taught at Schools. The book provides ample practice on various types of problems which can be solved by basic Algebra Formulae. This is the first "Algebra Practice Sets" volume of the series of books to be published in future.

These problems will provide an overall assessment of the student's progress in learning basic Algebra concepts and formulae taught in various secondary class textbooks. Students will definitely find this book useful in preparing for their examinations and evaluating their knowledge of Algebra.

This book also provides the method of solving these problems along with the answers which are provided at the end of this book. Students are encouraged to consciously apply their original thoughts in solving these problems on their own.

I look forward to constructive criticism which will help me in improving this book.

Sanjay Jamindar

Bangalore, India

CONTENTS

ALGEBRA BASIC FORMULAE

1. $(a+b)^2 = a^2 + 2ab + b^2$

2. $(a-b)^2 = a^2 - 2ab + b^2$

$\Rightarrow (a+b)^2 = (a-b)^2 + 4ab$

$\Rightarrow (a-b)^2 = (a+b)^2 - 4ab$

$\Rightarrow a^2 + b^2 = (a+b)^2 - 2ab = (a-b)^2 + 2ab = \dfrac{1}{2}\left\{(a+b)^2 + (a-b)^2\right\}$

$\Rightarrow ab = \left(\dfrac{a+b}{2}\right)^2 - \left(\dfrac{a-b}{2}\right)^2$

3. $(a+b)(a-b) = a^2 - b^2$

4. $(a+b)^3 = a^3 + 3a^2b + 3ab^2 + b^3$

$\Rightarrow a^3 + b^3 = (a+b)^3 - 3ab(a+b)$

5. $(a-b)^3 = a^3 - 3a^2b + 3ab^2 - b^3$

$\Rightarrow a^3 - b^3 = (a-b)^3 + 3ab(a-b)$

6. $a^3 + b^3 = (a+b)(a^2 - ab + b^2)$

7. $a^3 - b^3 = (a-b)(a^2 + ab + b^2)$

8. $(a+b+c)^2 = a^2 + b^2 + c^2 + 2(bc + ca + ab)$

9. $(a+b+c)^3 = a^3 + b^3 + c^3 + 3(a+b)(b+c)(c+a)$

10. $a^3 + b^3 + c^3 - 3abc = (a+b+c)(a^2 + b^2 + c^2 - ab - bc - ca)$

$= \dfrac{1}{2}(a+b+c)\left\{(a-b)^2 + (b-c)^2 + (c-a)^2\right\}$

$\Rightarrow a^2 + b^2 + c^2 - ab - bc - ca = \dfrac{1}{2}\left[(b-c)^2 + (c-a)^2 + (a-b)^2\right]$

PRACTICE SET 1

1. If $x + y = \dfrac{1}{x} + \dfrac{1}{y} = 100$, find the value of $x^3 + y^3$.

2. Express $(x+1)(x+2)(x+6)(x+7) + (x+4)^2$ as a perfect square.

3. If $ax + by = p$, $bx - ay = q$, $a^2 + b^2 = 1$, prove that $x^2 + y^2 = p^2 + q^2$.

4. If $x^4 + \dfrac{1}{x^4} = 119$, find the value of $x^3 - \dfrac{1}{x^3}$.

5. Simplify using formulae: $\left\{ (3a + 2b)^2 - (2a + b)^2 \right\}^2$.

6. Resolve into factors: $a^4 - 6a + 9 + 2a^3 - 5a^2$.

7. If $a + b + c = 9$, $ab + bc + ca = 26$, $a^3 + b^3 + c^3 = 138$, find the value of abc.

8. Resolve into factors: $(x+1)(x+3)(x-4)(x-6) + 24$.

9. If $2x - \dfrac{2}{x} = 3$, prove that $8\left(x^3 - \dfrac{1}{x^3} \right) = 63$.

10. If $x + y = \sqrt{3}$ & $x - y = \sqrt{2}$, find the value of $8xy(x^2 + y^2)$.

PRACTICE SET 2

1. If $a + b + c = 0$, prove that $a(b + c)^2 + b(c + a)^2 + c(a + b)^2 = 3abc$.

2. If $x = b - c$, $y = c - a$ & $z = a - b$, find the value of $x^2 + y^2 - z^2 + 2xy$.

3. Resolve into factors: $a^4 - 9a^2 + 30a - 25$.

4. If $x + y = a$, $x^2 + y^2 = b^2$ & $x^3 + y^3 = c^3$, prove that $a^3 + 2c^3 = 3ab^2$.

5. If $x = 2 + 2^{1/3} + 2^{2/3}$, prove that $x^3 - 6x^2 + 6x - 2 = 0$.

6. If $a + b + c = 6$, $ab + bc + ca = 11$,
   find the value of $bc(b + c) + ca(c + a) + ab(a + b) + 3abc$.

7. If $a = x + y$, $b = x - y$ & $c = x + 2y$, find the value of $a^2 + b^2 + c^2 - ab - bc - ca$.

8. If $x + y + z = 2$, $xy + yz + zx = 1$, find the value of $(x + y)^2 + (y + z)^2 + (z + x)^2$.

9. If $x + \dfrac{1}{x} = 5$, prove that $x^5 + \dfrac{1}{x^5} = 2525$.

10. If $p = 3 + \dfrac{1}{p}$, prove that $p^4 = 727 - \dfrac{1}{p^4}$.

PRACTICE SET 3

1. If $a^2 = ab + bc + ca - b^2 - c^2$, prove that $a^3 + b^3 + c^3 = 3abc$.

2. Multiply $\left(a^4 + a^2b^2 + b^4\right)$ & $\left(a^2 - b^2\right)$ using algebraic formulae.

3. If $2x - \dfrac{2}{x} = 3$, prove that $x^2 + \dfrac{1}{x^2} + 3 = 7\dfrac{1}{4}$.

4. Show that $n\left(n^2 + 5\right)$ is divisible by 6.

5. Resolve into factors: $4a^2 - 4ab + 2bc - c^2$.

6. If $x + y = 6$, $xy = 8$, find the value of $x^3 + y^3 + 4(x - y)^2$.

7. Express $(a + 2)(2a + 1)(5a + 2) - 3a^4$ as the difference of two squares.

8. Show that: $(a + b)^3 - (a - b)^3 = 3(a + b)^2(a - b) - 3(a + b)(a - b)^2 + 8b^3$.

9. If $a^8 + \dfrac{1}{a^8} = 2$, find the value of $a^{32} + \dfrac{1}{a^{32}}$.

10. If $x = 12$, prove that $x^5 - 13x^4 + 13x^3 - 13x^2 + 13x - 1 = 11$.

PRACTICE SET 4

1. If $x + \dfrac{2}{y} = 12$ and $12y - \dfrac{1}{z} = 2$, prove that $xyz = 1$.

2. If $a + \dfrac{1}{b} = b + \dfrac{1}{c} = c + \dfrac{1}{a}$, prove that $abc = \pm 1$.

3. Express $x(2x+1)(x-2)(2x-3) - 63$ as the difference of two squares.

4. If $m + \dfrac{1}{m} = 2$, show that $m^2 + \dfrac{1}{m^2} = m^3 + \dfrac{1}{m^3}$.

5. If $m - \dfrac{1}{m} = 2$, show that $m^4 + \dfrac{1}{m^4} = m^3 - \dfrac{1}{m^3} + 20$.

6. Find the value of $1.79 \times 1.79 + 2.42 \times 1.79 + 1.21 \times 1.21$.

7. If $y + mx = \sqrt{a^2 m^2 + b^2}$ and $x - my = -\sqrt{b^2 m^2 + a^2}$, show that $x^2 + y^2 = a^2 + b^2$.

8. Resolve into factors: $a^2 + b^2 + 2ab + 2a + 2b + 1$.

9. Find the value of $x^4 + y^4 - 2x^2 y^2$, when $x = a + \dfrac{1}{a}$, $y = a - \dfrac{1}{a}$.

10. If $bx = ay$, prove that $\left(x^2 + y^2\right)\left(a^2 + b^2\right) = (ax + by)^2$.

PRACTICE SET 5

1. If $x + y = 5$, $xy = 6$ and $x > y$, find the value of $4(x^2 - y^2) - x^3 + y^3$.

2. Express $(4a + 3b)(2a - 3b)$ as the difference of two square terms.

3. Resolve into factors: $a^{12} - b^{12}$.

4. If $x + \dfrac{1}{x} = 3$, find the value of $\dfrac{x}{x^2 + x + 1}$.

5. Multiply: $(x^2 + x + 1)$, $(x^2 - x + 1)$ & $(x^4 - x^2 + 1)$ using only algebraic formula $a^2 - b^2 = (a + b)(a - b)$.

6. Resolve into factors: $x^4 + 2x^3 - x^2 - 2x + 1$.

7. If $x - y = 4$, find the value of $x^3 - y^3 - 12xy$.

8. If $a = b + c$, prove that $a^3 - b^3 - c^3 = 3abc$.

9. If $x = 3^{\frac{1}{3}} - 3^{\frac{-1}{3}}$, prove that $3x^3 + 9x = 8$.

10. Express the following as difference of two squares:
$(x + 7)(x + 9)(x + 11)(x + 13)$

PRACTICE SET 6

1. If $a+b+c=0$, prove that $a^2 - bc = b^2 - ca = c^2 - ab$.

2. If $3(a^2 + b^2 + c^2) = (a+b+c)^2$, prove that $a = b = c$.

3. If $x^4 + \dfrac{1}{x^4} = 47$, find the values of $\left(x - \dfrac{1}{x}\right)^2$.

4. If $x+y+z=6$ & $xy+yz+zx=9$, prove that $\dfrac{1}{1-x} + \dfrac{1}{1-y} + \dfrac{1}{1-z} = 0$.

5. If $xy(x+y)=1$, prove that $\dfrac{1}{x^3 y^3} - x^3 - y^3 = 3$.

6. If $(x+y)=6, xy=5$, find the value of $x^2\left(\dfrac{x^2}{y} + y\right) + y^2\left(\dfrac{y^2}{x} + x\right)$.

7. Express $(x+2y)(x+3y)$ as difference of two squares.

8. If $a+b+c=2, ab+bc+ca=1$, find the value of $\dfrac{(a+b)^2 + (b+c)^2 + (c+a)^2}{a^2 + b^2 + c^2}$.

9. If $a-b=1$, $a+b=\sqrt{3}$, find the value of $a^4 - b^4$.

10. If $u = x + \dfrac{1}{x}$, show that $x^4 + \dfrac{1}{x^4} = u^4 - 4u^2 + 2$.

PRACTICE SET 7

1. Find out a, b when $a^2 + b^2 = 58$ & $ab = 21$.

2. If $x + y + z = a$, $x^2 + y^2 + z^2 = b$, find the value of $xy + yz + zx$.

3. If $a + b + c = 8$, $a^2 + b^2 + c^2 = 30$, find the value of $a^3 + b^3 + c^3 - 3abc$.

4. If $a = 1 - \dfrac{1}{b}$ and $b = \dfrac{2}{c} - 1$, prove that $a + b + c = 2 + \dfrac{b^3 - 1}{b(b+1)}$.

5. For what values of a and b will $27x^3 - 9ax^2 + 36x + b$ be a perfect cube?

6. Resolve into factors: $a^4 + 2a^3b - 2ab^3 - b^4$.

7. If $x = \sqrt[3]{5} - 5$, prove that $\dfrac{x^3 + 15x^2 + 75x + 225}{x^3 + 15x^2 + 75x + 119} = -105$.

8. If $x = b - c$, $y = c - a$, $z = a - b$, prove that $x^2 - y^2 + z^2 + 2zx = 0$.

9. If $x + y = 1 + xy$, show that $x^3 + y^3 = 1 + x^3 y^3$.

10. Resolve into factors: $\dfrac{a}{b} + \dfrac{b}{a} + \dfrac{c}{b} + \dfrac{b}{c} + \dfrac{a}{c} + \dfrac{c}{a} + 3$

PRACTICE SET 8

1. If $x + \dfrac{4}{x} = 4$, find the value of $x^3 + \dfrac{4}{x^3}$.

2. If $x - \dfrac{1}{x} = -3$, then find the value of $x^4 + \dfrac{1}{x^4}$.

3. Simplify: $\dfrac{1}{1-x} + \dfrac{1}{1+x} + \dfrac{2}{1+x^2} + \dfrac{4}{1+x^4} - \dfrac{8}{1-x^8}$

4. Resolve into factors: $\left(a^2 - b^2\right)\left(x^2 + y^2\right) + 2\left(a^2 + b^2\right)xy$

5. If $\left(a + \dfrac{1}{a}\right)^2 = 3$, then prove that $a^3 + \dfrac{1}{a^3} = 0$.

6. Prove that $(2x-1)^3 - (x-2)^3 - 3(2x-1)(x-2)(x+1) = (x+1)^3$.

7. If $\left(x - \dfrac{1}{x}\right)^2 = 3$, find the value of $x^6 + \dfrac{1}{x^6}$.

8. If $a + b = 1$, prove that $\left(a^2 - b^2\right)^2 = a^3 + b^3 - ab$.

9. If $a^2 + b^2 + c^2 = 9$ & $ab + bc + ca = 8$, find the value of $(a+b+c)$.

10. If $x - y = 4z$, then find the value of $\dfrac{x^3 - y^3 - 12xyz}{(x-y)^2}$.

PRACTICE SET 9

1. If $x = b + c - a$, $y = c + a - b$, $z = a + b - c$, show that
   $$x^3 + y^3 + z^3 - 3xyz = 4\left(a^3 + b^3 + c^3 - 3abc\right).$$

2. Resolve into factors: $(a - 1)x^2 + a^2 xy + (a + 1)y^2$.

3. Simplify: $(x + 4)\left(x^2 - 4x + 16\right) - (x + 3)\left(x^2 - 3x + 9\right)$.

4. Resolve into factors: $x^3 - y^3 + 3y^2 - 3y + 1$.

5. If $\left(x - \dfrac{1}{x}\right)^2 = 3$, find the value of $x^3 + \dfrac{1}{x^3}$.

6. Resolve into factors: $\left(a^2 - b^2\right)\left(x^2 - y^2\right) + 4abxy$.

7. If $\left(a^{1/3} + b^{1/3} + c^{1/3}\right) = 0$ & $abc = 1$, find the value of $a + b + c$.

8. Simplify: $(y + z)(y - z)\left(y^2 + z^2\right) + (z + x)(z - x)\left(z^2 + x^2\right) + (x + y)(x - y)\left(x^2 + y^2\right)$.

9. If $x + y = 99$, find the value of $x^3 + y^3 + 297xy$.

10. Express $x(x - 3)(x - 6)(x - 9) + 81$ as a square term.

PRACTICE SET 10

1. Resolve into factors: $(x+1)(x+3)(x+5)(x+7)+15$.

2. Simplify: $(x+y+z)(x-y+z)(x+y-z)(z+y-x)$.

3. If $7a+\dfrac{1}{4a}=-2$, find the value of

$$\dfrac{1}{16a^2}+\dfrac{(a-b)^2+(b-c)^2+(c-a)^2}{(a-b)(b-c)(c-a)}+\dfrac{2}{a-b}+\dfrac{2}{b-c}+\dfrac{2}{c-a}+49a^2.$$

4. If $x+y=2a$ & $y=a-\dfrac{1}{a}$, find the value of $\dfrac{x^4+y^4-2x^2y^2}{(x+y)^3}$.

5. If $a+b=3$ & $ab=2$, find the value of a^6+b^6.

6. Express $(x+1)(x+3)(x-5)(x-7)$ as difference of two squares.

7. If $x=\sqrt{3}-\dfrac{1}{\sqrt{3}}$ & $y=\sqrt{3}+\dfrac{1}{\sqrt{3}}$, prove that $\dfrac{x^2}{y}+\dfrac{y^2}{x}=3\sqrt{3}$.

8. Resolve into factors: $(x^2-1)(x+2)x-8$.

9. If $\left(x-\dfrac{1}{x}\right)^2=3$, then find the value of $x^6+\dfrac{1}{x^6}$.

10. Resolve into factors: $(x+1)(x+2)(3x-1)(3x-4)+12$.

ANSWERS

PRACTICE SET 1:

1. 999700
2. $\left(x^2 + 8x + 10\right)^2$
3. Check Solution.
4. 36
5. $\left(5a^2 + 8ab + 3b^2\right)^2$
6. $\left(a^2 + a - 3\right)\left(a^2 + a - 3\right)$
7. 37
8. $\left(x^2 - 3x - 6\right)\left(x^2 - 3x - 16\right)$
9. Check Solution.
10. 5

PRACTICE SET 2:

1. Check Solution.
2. 0
3. $\left(a^2 + 3a - 5\right)\left(a^2 - 3a + 5\right)$
4. Check Solution.
5. Check Solution.
6. 66
7. $7y^2$
8. 6
9. Check Solution.
10. Check Solution.

PRACTICE SET 3:

1. Check Solution.
2. $a^6 - b^6$
3. Check Solution.
4. Check Solution.
5. $(2a - c)(2a - 2b + c)$
6. 88
7. $(a^2 + x)^2 - (2a^2)^2$
8. Check Solution.
9. 2
10. Check Solution.

PRACTICE SET 4:

1. Check Solution.
2. Check Solution.
3. $(2x^2 - 3x - 1)^2 - 8^2$
4. Check Solution.
5. Check Solution.
6. 9
7. Check Solution.
8. $(a + b + 1)(a + b + 1)$
9. 16
10. Check Solution.

PRACTICE SET 5:

1. 1
2. $(3a)^2 - (a + 3b)^2$
3. $(a^2 + b^2)(a^4 - a^2b^2 + b^4)(a + b)(a^2 - ab + b^2)(a - b)(a^2 + ab + b^2)$
4. $\dfrac{1}{4}$
5. $x^8 + x^4 + 1$

6. $(x^2 + x - 1)(x^2 + x - 1)$
7. 64
8. Check Solution.
9. Check Solution.
10. $(x^2 + 20x + 95)^2 - 4^2$

PRACTICE SET 6:

1. Check Solution.
2. Check Solution.
3. $5, -9$
4. Check Solution.
5. Check Solution.
6. 655.2
7. $\left\{ \dfrac{1}{2}(2x + 5y) \right\}^2 - \left(\dfrac{1}{2}y \right)^2$
8. 3
9. $2\sqrt{3}$
10. Check Solution.

PRACTICE SET 7:

1. $a = 7$, $b = 3$; $a = 3$, $b = 7$; $a = -7$, $b = -3$; $a = -3$, $b = -7$.
2. $\dfrac{1}{2}(a^2 - b)$
3. 104
4. Check Solution.
5. $a = 6, -6$; $b = -8, 8$
6. $(a - b)(a + b)(a + b)(a + b)$
7. Check Solution.
8. Check Solution.
9. Check Solution.

10. $\left(a+b+c\right)\left(\dfrac{1}{a}+\dfrac{1}{b}+\dfrac{1}{c}\right)$

PRACTICE SET 8:

1. $8\dfrac{1}{2}$
2. 119
3. 0
4. $\left(ax+ay+bx-by\right)\left(ax+ay-bx+by\right)$
5. Check Solution.
6. Check Solution.
7. 110
8. Check Solution.
9. ± 5
10. $4z$

PRACTICE SET 9:

1. Check Solution.
2. $\left(x+ay+y\right)\left(ax-x+y\right)$
3. 37
4. $\left(x-y+1\right)\left(x^2+y^2+1+xy-x-2y\right)$
5. $4\sqrt{7}$
6. $\left(ax+by+ay-bx\right)\left(ax+by-ay+bx\right)$
7. 3
8. 0
9. 970299
10. $\left(x^2-9x+9\right)^2$

PRACTICE SET 10:

1. $(x+2)(x+6)(x^2+8x+10)$

2. $2x^2y^2 + 2y^2z^2 + 2z^2x^2 - x^4 - y^4 - z^4$

3. $\dfrac{1}{2}$

4. $\dfrac{2}{a^3}$

5. 65

6. $(x^2-4x-13)^2 - 8^2$

7. Check Solution.

8. $(x^2+x+2)(x^2+x-4)$

9. 110

10. $(x-1)(3x+5)(3x^2+2x-4)$

SOLUTIONS

PRACTICE SET 1:

Problem 1:

$\dfrac{1}{x} + \dfrac{1}{y} = 100$

$\Rightarrow x + y = 100xy$

$\Rightarrow 100 = 100xy \quad [\Theta\ x + y = 100]$

$\Rightarrow xy = 1$

$\therefore x^3 + y^3 = (x+y)^3 - 3xy(x+y) = 1000000 - 3.1.100 = 999700$ **(Ans.)**

Problem 2:

$(x+1)(x+2)(x+6)(x+7) + (x+4)^2$

$= (x+1)(x+7)(x+2)(x+6) + (x+4)^2$ [Rearranging terms]

$= (x^2 + 8x + 7)(x^2 + 8x + 12) + (x^2 + 8x + 16)$

$= (a+7)(a+12) + (a+16)$ [Assume $a = x^2 + 8x$]

$= a^2 + 19a + 84 + a + 16$

$= a^2 + 20a + 100$

$= (a+10)^2 \quad [\Theta\ (a+b)^2 = a^2 + 2ab + b^2]$

$= (x^2 + 8x + 10)^2$ **(Ans.)**

Problem 3:

$ax + by = p$

$\Rightarrow (ax + by)^2 = p^2$

$\Rightarrow a^2x^2 + 2abxy + b^2y^2 = p^2 \quad .....(1)$

$bx - ay = q$

$$\Rightarrow (bx - ay)^2 = q^2$$
$$\Rightarrow b^2 x^2 - 2abxy + a^2 y^2 = q^2 \quad .....(2)$$

Adding (1) and (2)
$$a^2 x^2 + b^2 x^2 + b^2 y^2 + a^2 y^2 = p^2 + q^2$$
$$\Rightarrow x^2 (a^2 + b^2) + y^2 (a^2 + b^2) = p^2 + q^2$$
$$\Rightarrow x^2 + y^2 = p^2 + q^2 \text{ (Ans.)}$$
$$[\Theta \ a^2 + b^2 = 1]$$

Problem 4:

$$x^4 + \frac{1}{x^4} = 119$$

$$\Rightarrow \left(x^2 + \frac{1}{x^2} \right)^2 - 2.x^2 . \frac{1}{x^2} = 119$$

$$\Rightarrow \left(x^2 + \frac{1}{x^2} \right)^2 = 121 \qquad \Rightarrow x^2 + \frac{1}{x^2} = 11$$

$$\left(x - \frac{1}{x} \right)^2 = x^2 + \frac{1}{x^2} - 2.x.\frac{1}{x} = 11 - 2 = 9 \qquad \Rightarrow x - \frac{1}{x} = 3$$

$$\therefore \ x^3 - \frac{1}{x^3} = \left(x - \frac{1}{x} \right)^3 + 3.x.\frac{1}{x} \left(x - \frac{1}{x} \right) \quad [\Theta \ a^3 - b^3 = (a - b)^3 + 3ab(a - b)]$$

$$= 27 + 3.3 = 36 \text{ (Ans.)}$$

Problem 5:

Assume $x = (3a + 2b), \ y = (2a + b)$
$$\therefore \left\{ (3a + 2b)^2 - (2a + b)^2 \right\}^2$$
$$= \left(x^2 - y^2 \right)^2$$
$$= (x + y)^2 (x - y)^2 \quad [\Theta \ (a + b)(a - b) = a^2 - b^2]$$
$$= (5a + 3b)^2 (a + b)^2$$
$$= \left(5a^2 + 8ab + 3b^2 \right)^2 \text{ (Ans.)}$$

Problem 6:

$$a^4 - 6a + 9 + 2a^3 - 5a^2$$
$$= a^4 + \left(a^2 - 6a + 9\right) + 2a^3 - 6a^2 \quad \text{[Rearranging terms]}$$
$$= \left(a^2\right)^2 + (a-3)^2 + 2a^2(a-3)$$
$$= x^2 + y^2 + 2xy \quad \text{[Assume } x = a^2, \ y = a-3\text{]}$$
$$= (x+y)^2 = \left(a^2 + a - 3\right)\left(a^2 + a - 3\right) \quad \textbf{(Ans.)}$$

Problem 7:

$$a^2 + b^2 + c^2 = (a+b+c)^2 - 2(ab+bc+ca) = 81 - 2.26 = 29$$
$$[\ \Theta\ a+b+c = 9\ \&\ ab+bc+ca = 26\ ]$$
$$\therefore a^3 + b^3 + c^3 - 3abc = (a+b+c)\left(a^2 + b^2 + c^2 - bc - ca - ab\right)$$
$$\Rightarrow 138 - 3abc = 9.(29 - 26) = 9.3 = 27 \quad [\ \Theta\ a^3 + b^3 + c^3 = 138\ ]$$
$$\Rightarrow abc = 37 \quad \textbf{(Ans.)}$$

Problem 8:

$$(x+1)(x+3)(x-4)(x-6) + 24$$
$$= (x+1)(x-4)(x+3)(x-6) + 24 \quad \text{[Rearranging terms]}$$
$$= \left(x^2 - 3x - 4\right)\left(x^2 - 3x - 18\right) + 24$$
$$= (a-4)(a-18) + 24 \quad \text{[Assume } a = x^2 - 3x\text{]}$$
$$= a^2 - 22a + 96$$
$$= a^2 - 16a - 6a + 96 \quad \text{[Breaking the middle term]}$$
$$= (a-6)(a-16) \quad [\ \Theta\ x^2 + (a+b)x + ab = (x+a)(x+b)\ ]$$
$$= \left(x^2 - 3x - 6\right)\left(x^2 - 3x - 16\right) \quad \textbf{(Ans.)}$$

Problem 9:

$$8\left(x^3 - \frac{1}{x^3}\right) = (2x)^3 - \left(\frac{2}{x}\right)^3$$
$$= \left(2x - \frac{2}{x}\right)^3 + 3.2x.\frac{2}{x}\left(2x - \frac{2}{x}\right) \quad [\ \Theta\ a^3 - b^3 = (a-b)^3 + 3ab(a-b)\ ]$$
$$= 27 + 12.3 = 63 \quad \textbf{(Ans.)}$$

Problem 10:

$x + y = \sqrt{3} \Rightarrow (x+y)^2 = 3 \Rightarrow x^2 + y^2 + 2xy = 3$ ..... (1)

$x - y = \sqrt{2} \Rightarrow (x-y)^2 = 2 \Rightarrow x^2 + y^2 - 2xy = 2$ .....(2)

Adding (1) and (2)

$2(x^2 + y^2) = 5$

Subtracting (1) and (2)

$4xy = 1$

$\therefore 8xy(x^2 + y^2) = 2.\dfrac{5}{2} = 5$ **(Ans.)**

PRACTICE SET 2:

Problem 1:

$\Theta\ a + b + c = 0$

$\Rightarrow a + b = -c,\ b + c = -a,\ c + a = -b$

$\therefore a(b+c)^2 + b(c+a)^2 + c(a+b)^2 = a^3 + b^3 + c^3 = 3abc$ **(Ans.)**

$[\ \Theta\ a + b + c = 0$ and $a^3 + b^3 + c^3 - 3abc = (a+b+c)(a^2 + b^2 + c^2 - ab - bc - ca)]$

Problem 2:

$\Theta\ x = b - c,\ y = c - a,\ z = a - b$

$\therefore x + y + z = 0$

$\therefore x^2 + y^2 - z^2 + 2xy = (x+y)^2 - z^2 = (x+y+z)(x+y-z) = 0$ **(Ans.)**

Problem 3:

$a^4 - 9a^2 + 30a - 25$

$= a^4 - (9a^2 - 30a + 25)$

$= (a^2)^2 - (3a - 5)^2$

$= (a^2 + 3a - 5)(a^2 - 3a + 5)$ **(Ans.)**

$[\Theta\ (a+b)(a-b) = a^2 - b^2]$

Problem 4:

$a^3 + 2c^3$

$= (x+y)^3 + 2(x^3 + y^3) \quad [\Theta\; x+y = a ,\; x^3 + y^3 = c^3]$

$= x^3 + y^3 + 3xy(x+y) + 2x^3 + 2y^3 \quad [\Theta\; (a+b)^3 = a^3 + 3a^2b + 3ab^2 + b^3]$

$= 3(x^3 + y^3) + 3xy(x+y)$

$= 3(x+y)(x^2 - xy + y^2) + 3xy(x+y) \quad [\Theta\; a^3 + b^3 = (a+b)(a^2 - ab + b^2)]$

$= 3(x+y)(x^2 + y^2)$

$= 3ab^2$ **(Ans.)**

$[\Theta\; x^2 + y^2 = b^2]$

Problem 5:

$x = 2 + 2^{1/3} + 2^{2/3}$

$\Rightarrow x - 2 = 2^{1/3} + 2^{2/3}$

$\Rightarrow (x-2)^3 = (2^{1/3} + 2^{2/3})^3$

$\Rightarrow x^3 - 3x^2.2 + 3x.2^2 - 8 = \left(2^{\frac{1}{3}}\right)^3 + \left(2^{\frac{2}{3}}\right)^3 + 3.2^{\frac{1}{3}}.2^{\frac{2}{3}}\left(2^{\frac{1}{3}} + 2^{\frac{2}{3}}\right)$

$\Rightarrow x^3 - 6x^2 + 12x - 8 = 2 + 2^2 + 6(x-2) \quad [\Theta\; x - 2 = 2^{1/3} + 2^{2/3}]$

$\Rightarrow x^3 - 6x^2 + 6x - 2 = 0$ **(Ans.)**

Problem 6:

$bc(b+c) + ca(c+a) + ab(a+b) + 3abc$

$= \{bc(b+c) + abc\} + \{ca(c+a) + abc\} + \{ab(a+b) + abc\}$

$= bc(a+b+c) + ca(a+b+c) + ab(a+b+c)$

$= (a+b+c)(bc + ca + ab) = 6.11 = 66$ **(Ans.)**

$[\Theta\; a+b+c = 6,\; ab+bc+ca = 11]$

Problem 7:

$a^2 + b^2 + c^2 - ab - bc - ca$

$= \frac{1}{2}\{(a-b)^2 + (b-c)^2 + (c-a)^2\}$ [using formula directly]

$= \frac{1}{2}(4y^2 + 9y^2 + y^2) = 7y^2$ **(Ans.)**

$[\Theta\; a = x+y,\; b = x-y,\; c = x+2y]$

Problem 8:

$$(x+y)^2 + (y+z)^2 + (z+x)^2$$
$$= 2(x^2 + y^2 + z^2 + xy + yz + zx)$$
$$= 2\{(x^2 + y^2 + z^2) + 2(xy + yz + zx) - (xy + yz + zx)\}$$
$$= 2\{(x + y + z)^2 - (xy + yz + zx)\} \quad [\Theta\ (a+b+c)^2 = a^2 + b^2 + c^2 + 2(bc + ca + ab)]$$
$$= 2(2^2 - 1) = 6 \ \textbf{(Ans.)}$$
$$[\Theta\ x + y + z = 2,\ xy + yz + zx = 1]$$

Problem 9:

$$x^2 + \frac{1}{x^2} = \left(x + \frac{1}{x}\right)^2 - 2.x.\frac{1}{x} = 25 - 2 = 23 \quad [\Theta\ x + \frac{1}{x} = 5]$$

$$x^3 + \frac{1}{x^3} = \left(x + \frac{1}{x}\right)^3 - 3.x.\frac{1}{x}\left(x + \frac{1}{x}\right) = 125 - 15 = 110 \quad [\Theta\ x + \frac{1}{x} = 5]$$

$$\therefore \left(x^2 + \frac{1}{x^2}\right)\left(x^3 + \frac{1}{x^3}\right) = 23 \times 110 = 2530$$

$$\Rightarrow x^5 + \frac{1}{x} + x + \frac{1}{x^5} = 2530$$

$$\Rightarrow x^5 + \frac{1}{x^5} + \left(x + \frac{1}{x}\right) = 2530$$

$$\therefore x^5 + \frac{1}{x^5} = 2530 - 5 = 2525 \ \textbf{(Ans.)}$$

Problem 10:

$$p = 5 + \frac{1}{p}$$

$$\Rightarrow \left(p - \frac{1}{p}\right)^2 = 25$$

$$\Rightarrow p^2 + \frac{1}{p^2} - 2.p.\frac{1}{p} = 25 \quad [\Theta\ (a-b)^2 = a^2 - 2ab + b^2]$$

$$\Rightarrow p^2 + \frac{1}{p^2} = 27$$

$$\Rightarrow \left(p^2 + \frac{1}{p^2} \right)^2 = 729 \quad [\Theta\ (a+b)^2 = a^2 + 2ab + b^2]$$

$$\Rightarrow p^4 = 727 - \frac{1}{p^4} \text{ (Ans.)}$$

PRACTICE SET 3:

Problem 1:
$$a^3 + b^3 + c^3 - 3abc = (a+b+c)(a^2 + b^2 + c^2 - ab - bc - ca) = 0$$
$$[\Theta\ a^2 = ab + bc + ca - b^2 - c^2]$$
$$\Rightarrow a^3 + b^3 + c^3 = 3abc \text{ (Ans.)}$$

Problem 2:
$$\left(a^4 + a^2 b^2 + b^4 \right)\left(a^2 - b^2 \right)$$
$$= \left\{ \left(a^2 \right)^2 + a^2 b^2 + \left(b^2 \right)^2 \right\}\left(a^2 - b^2 \right) = \left(a^2 \right)^3 - \left(b^2 \right)^3 \quad [\Theta\ a^3 - b^3 = (a-b)(a^2 + ab + b^2)]$$
$$= a^6 - b^6 \text{ (Ans.)}$$

Problem 3:
$$2x - \frac{2}{x} = 3$$
$$\Rightarrow x - \frac{1}{x} = \frac{3}{2}$$
$$\therefore x^2 + \frac{1}{x^2} + 3 = \left(x - \frac{1}{x} \right)^2 + 2.x.\frac{1}{x} + 3 = \frac{9}{4} + 5 = 7\frac{1}{4} \text{ (Ans.)}$$

Problem 4:
$$n\left(n^2 + 5 \right)$$
$$= n^3 + 5n$$
$$= n^3 - n + 6n$$
$$= n\left(n^2 - 1 \right) + 6n = n(n-1)(n+1) + 6n$$

Now, for any integral value for n, since $(n-1)$, n and $(n+1)$ are three consecutive integers, their product is divisible by 6. $6n$ is also divisible by 6.

Problem 5:

$$4a^2 - 4ab + 2bc - c^2$$
$$= 4a^2 - 4ab + b^2 - b^2 + 2bc - c^2$$
$$= \left(4a^2 - 4ab + b^2\right) - \left(b^2 - 2bc + c^2\right)$$
$$= (2a - b)^2 - (b - c)^2$$
$$= (2a - b + b - c)(2a - b - b + c) = (2a - c)(2a - 2b + c) \textbf{ (Ans.)}$$

Problem 6:

$$x^3 + y^3 + 4(x - y)^2$$
$$= (x + y)^3 - 3xy(x + y) + 4\left\{(x + y)^2 - 4xy\right\} \quad \text{[Applying formula directly]}$$
$$= 88 \textbf{ (Ans.)}$$
$$[\Theta \; x + y = 6, xy = 8]$$

Problem 7:

$$(a + 2)(2a + 1)(5a + 2) - 3a^4$$
$$= \left(2a^2 + 5a + 2\right)(5a + 2) - 3a^4$$
$$= \left(2a^2 + x\right)(x) - 3a^4 \quad \text{[Suppose } 5a + 2 = x]$$
$$= \left(x^2 + 2a^2 x + a^4\right) - 4a^4 = \left(a^2 + x\right)^2 - \left(2a^2\right)^2 \textbf{ (Ans.)}$$

Problem 8:

$$(a + b)^3 - 3(a + b)^2(a - b) + 3(a + b)(a - b)^2 - (a - b)^3$$
$$= x^3 - 3x^2 y + 3xy^2 - y^3 \quad \text{[Assuming } x = a + b, \; y = a - b]$$
$$= (x - y)^3 = (a + b - a + b)^3 = 8b^3$$
$$\Rightarrow (a + b)^3 - (a - b)^3 = 3(a + b)^2(a - b) - 3(a + b)(a - b)^2 + 8b^3 \textbf{ (Ans.)}$$

Problem 9:

$$a^{32} + \frac{1}{a^{32}}$$

$$= \left(a^{16}\right)^2 + \left(\frac{1}{a^{16}}\right)^2 = \left(a^{16} + \frac{1}{a^{16}}\right)^2 - 2.a^2.\frac{1}{a^2} \quad [\Theta\, a^2 + b^2 = (a+b)^2 - 2ab]$$

$$= \left\{\left(a^8\right)^2 + \frac{1}{\left(a^8\right)^2}\right\}^2 - 2 = \left\{\left(a^8 + \frac{1}{a^8}\right)^2 - 2\right\}^2 - 2 = (4-2)^2 - 2 = 2 \text{ (Ans.)}$$

$$[\Theta\, a^8 + \frac{1}{a^8} = 2]$$

Problem 10:

$$x^5 - 13x^4 + 13x^3 - 13x^2 + 13x - 1$$
$$= x^5 - 12x^4 - x^4 + 12x^3 + x^3 - 12x^2 - x^2 + 12x + x - 1$$
$$= x^4(x-12) - x^3(x-12) + x^2(x-12) - x(x-12) + (x-1)$$
$$= (x-1) \quad [\Theta\, x - 12 = 0 \text{ as } x = 12]$$
$$= 11 \text{ (Ans.)}$$

PRACTICE SET 4:

Problem 1:

$$x + \frac{2}{y} = 12$$
$$\Rightarrow xy = 12y - 2$$
$$\Rightarrow xy = 12y - \left(12y - \frac{1}{z}\right) \quad [\Theta\, 12y - \frac{1}{z} = 2]$$
$$\Rightarrow xy = \frac{1}{z} \quad \Rightarrow xyz = 1 \text{ (Ans.)}$$

Problem 2:

$$a + \frac{1}{b} = b + \frac{1}{c} \quad \Rightarrow a - b = \frac{1}{c} - \frac{1}{b} = \frac{b-c}{bc} \quad ....(1)$$

$$b + \frac{1}{c} = c + \frac{1}{a} \quad \Rightarrow b - c = \frac{1}{a} - \frac{1}{c} = \frac{c-a}{ac} \quad ....(2)$$

$$a + \frac{1}{b} = c + \frac{1}{a} \quad \Rightarrow c - a = \frac{1}{b} - \frac{1}{a} = \frac{a-b}{ab} \quad ....(3)$$

Multiplying (1), (2) and (3)

$$(a-b)(b-c)(c-a) = \frac{(b-c)(c-a)(a-b)}{a^2 b^2 c^2}$$

$$\Rightarrow a^2 b^2 c^2 = 1 \quad \Rightarrow abc = \pm 1 \quad \textbf{(Ans.)}$$

Problem 3:

$$x(2x+1)(x-2)(2x-3) - 63$$

$$= \{x(2x-3)\}\{(2x+1)(x-2)\} - 63$$

$$= (2x^2 - 3x)(2x^2 - 3x - 2) - 63$$

$$= a(a-2) - 63 \text{ [Assume } (2x^2 - 3x) = a]$$

$$= a^2 - 2a - 63 = a^2 - 2a + 1 - 1 - 63 = (a-1)^2 - 8^2 = (2x^2 - 3x - 1)^2 - 8^2 \quad \textbf{(Ans.)}$$

Problem 4:

$$m + \frac{1}{m} = 2$$

$$\therefore m^2 + \frac{1}{m^2} = \left(m + \frac{1}{m}\right)^2 - 2.m.\frac{1}{m} = 4 - 2 = 2$$

$$\therefore m^3 + \frac{1}{m^3} = \left(m + \frac{1}{m}\right)^3 - 3.m.\frac{1}{m}\left(m + \frac{1}{m}\right) = 2^3 - 3.2 = 2$$

$$\therefore m^2 + \frac{1}{m^2} = m^3 + \frac{1}{m^3} \quad \textbf{(Ans.)}$$

Problem 5:

$$m^3 - \frac{1}{m^3} + 20 = \left(m - \frac{1}{m}\right)^3 + 3.m.\frac{1}{m}\left(m - \frac{1}{m}\right) + 20 = 34 \quad [\Theta \ m - \frac{1}{m} = 2]$$

$$m^4 + \frac{1}{m^4} = (m^2)^2 + \left(\frac{1}{m^2}\right)^2 = \left(m^2 + \frac{1}{m^2}\right)^2 - 2.m^2.\frac{1}{m^2}$$

$$= \left\{\left(m - \frac{1}{m}\right)^2 + 2.m.\frac{1}{m}\right\}^2 - 2 = 6^2 - 2 = 34 \quad [\Theta \ m - \frac{1}{m} = 2]$$

$$\therefore m^4 + \frac{1}{m^4} = m^3 - \frac{1}{m^3} + 20 \text{ (Ans.)}$$

Problem 6:

Assume $1.79 = a$, $1.21 = b$

$$\therefore a + b = 1.79 + 1.21 = 3$$

$$\therefore 1.79 \times 1.79 + 2.42 \times 1.79 + 1.21 \times 1.21$$

$$= (1.79)^2 + 2 \times 1.21 \times 1.79 + (1.21)^2$$

$$= a^2 + 2ab + b^2 = (a+b)^2 = 9 \text{ (Ans.)}$$

Problem 7:

$$y + mx = \sqrt{a^2 m^2 + b^2}$$

$$\Rightarrow y^2 + m^2 x^2 + 2ymx = a^2 m^2 + b^2 \ \ldots (1) \ \text{[Squaring both sides]}$$

$$x - my = -\sqrt{b^2 m^2 + a^2}$$

$$\Rightarrow x^2 + m^2 y^2 - 2ymx = b^2 m^2 + a^2 \ \ldots (2) \ \text{[Squaring both sides]}$$

Adding (1) and (2),

$$\left(x^2 + y^2\right)\left(m^2 + 1\right) = \left(a^2 + b^2\right)\left(m^2 + 1\right)$$

$$\Rightarrow x^2 + y^2 = a^2 + b^2 \text{ (Ans.)}$$

Problem 8:

$$a^2 + b^2 + 2ab + 2a + 2b + 1$$

$$= \left(a^2 + b^2 + 2ab\right) + 2(a+b) + 1$$

$$= (a+b+1)^2 = (a+b+1)(a+b+1) \text{ (Ans.)}$$

Problem 9:

$$x^4 + y^4 - 2x^2 y^2$$

$$= \left(x^2\right)^2 - 2x^2 y^2 + \left(y^2\right)^2$$

$$= \left(x^2 - y^2\right)^2 = (x+y)^2 (x-y)^2 = (2a)^2 . \left(\frac{2}{a}\right)^2 = 16 \text{ (Ans.)}$$

$$[\Theta \ x = a + \frac{1}{a}, \ y = a - \frac{1}{a}]$$

Problem 10:

$$\left(x^2 + y^2\right)\!\left(a^2 + b^2\right) = x^2a^2 + x^2b^2 + y^2a^2 + y^2b^2$$

$$= x^2a^2 + xb.xb + ya.ya + y^2b^2$$

$$= x^2a^2 + xb.ya + xb.ya + y^2b^2 \quad [\Theta\, bx = ay]$$

$$= x^2a^2 + xa.yb + xa.yb + y^2b^2 = (xa)^2 + 2.xa.yb + (yb)^2 = (ax + by)^2 \;\; \textbf{(Ans.)}$$

PRACTICE SET 5:

Problem 1:

$$(x - y)^2 = (x + y)^2 - 4xy = 1 \;\; [\Theta\, x + y = 5, xy = 6]$$

$$\therefore x - y = 1$$

$$x^3 - y^3 = (x - y)^3 + 3xy(x - y) = 19$$

$$\therefore 4\!\left(x^2 - y^2\right) - x^3 + y^3$$

$$= 4(x + y)(x - y) - \left(x^3 - y^3\right)$$

$$= 4 \times 5 \times 1 - 19 = 1 \;\; \textbf{(Ans.)}$$

Problem 2:

$$\Theta\, ab = \left(\frac{a + b}{2}\right)^2 - \left(\frac{a - b}{2}\right)^2$$

$$\therefore (4a + 3b)(2a - 3b) = \left(\frac{(4a + 3b) + (2a - 3b)}{2}\right)^2 - \left(\frac{(4a + 3b) - (2a - 3b)}{2}\right)^2$$

$$= (3a)^2 - (a + 3b)^2 \;\; \textbf{(Ans.)}$$

Problem 3:

$$a^{12} - b^{12}$$

$$= \left(a^6\right)^2 - \left(b^6\right)^2$$

$$= \left(a^6 + b^6\right)\!\left(a^6 - b^6\right)$$

$$= \left\{\left(a^2\right)^3 + \left(b^2\right)^3\right\}\!\left\{\left(a^3\right)^2 - \left(b^3\right)^2\right\}$$

$$= \left(a^2 + b^2\right)\left(\left(a^2\right)^2 - a^2.b^2 + \left(b^2\right)^2\right)\left(a^3 + b^3\right)\left(a^3 - b^3\right)$$
$$= \left(a^2 + b^2\right)\left(a^4 - a^2b^2 + b^4\right)(a+b)\left(a^2 - ab + b^2\right)(a-b)\left(a^2 + ab + b^2\right) \textbf{ (Ans.)}$$

Problem 4:

$$x + \frac{1}{x} = 3 \Rightarrow x^2 + 1 = 3x$$

$$\therefore \frac{x}{x^2 + x + 1} = \frac{x}{3x + x} = \frac{1}{4} \textbf{ (Ans.)}$$

Problem 5:

$$\left(x^2 + x + 1\right)\left(x^2 - x + 1\right)\left(x^4 - x^2 + 1\right)$$
$$= \left\{\left(x^2 + 1\right) + x\right\}\left\{\left(x^2 + 1\right) - x\right\}\left(x^4 - x^2 + 1\right)$$
$$= \left\{\left(x^2 + 1\right)^2 - x^2\right\}\left(x^4 - x^2 + 1\right) = \left\{\left(x^4 + 1\right) + x^2\right\}\left\{\left(x^4 + 1\right) - x^2\right\} = \left(x^4 + 1\right)^2 - \left(x^2\right)^2$$
$$= x^8 + x^2 + 1 \textbf{ (Ans.)}$$

Problem 6:

$$x^4 + 2x^3 - x^2 - 2x + 1$$
$$= x^4 + 2x^3 + x^2 - 2x^2 - 2x + 1$$
$$= \left(x^2 + x\right)^2 - 2.\left(x^2 + x\right)1 + (1)^2$$
$$= a^2 - 2a + 1^2 \text{ [Assume } x^2 + x = a\text{]}$$
$$= (a-1)^2 = \left(x^2 + x - 1\right)\left(x^2 + x - 1\right) \textbf{ (Ans.)}$$

Problem 7:

$$x^3 - y^3 - 12xy = (x-y)^3 + 3xy(x-y) - 12xy = 4^3 + 3xy.4 - 12xy = 64 \textbf{ (Ans.)}$$

Problem 8:

$$a^3 - b^3 - c^3 = a^3 - \left(b^3 + c^3\right)$$
$$= a^3 - \left\{(b+c)^3 - 3bc(b+c)\right\} = a^3 - \left(a^3 - 3bc.a\right) = 3abc \textbf{ (Ans.)}$$
$$[\Theta \, b + c = a]$$

Problem 9:

$x = 3^{\frac{1}{3}} - 3^{-\frac{1}{3}}$

$\Rightarrow x^3 = \left(3^{\frac{1}{3}} - 3^{-\frac{1}{3}}\right)^3 = \left(3^{\frac{1}{3}}\right)^3 - \left(3^{-\frac{1}{3}}\right)^3 - 3.3^{\frac{1}{3}}.3^{-\frac{1}{3}}\left(3^{\frac{1}{3}} - 3^{-\frac{1}{3}}\right)$

$= 3 - \dfrac{1}{3} - 3x = \dfrac{8}{3} - 3x \Rightarrow 3x^3 + 9x = 8$ **(Ans.)**

Problem 10:

$(x+7)(x+9)(x+11)(x+13)$

$= \{(x+7)(x+13)\}\{(x+9)(x+11)\}$

$= \left(x^2 + 20x + 91\right)\left(x^2 + 20x + 99\right) = (a+91)(a+99)\,[\text{Assume } a = x^2 + 20x]$

$= \left\{\dfrac{(a+99)+(a+91)}{2}\right\}^2 - \left\{\dfrac{(a+99)-(a+91)}{2}\right\}^2 = (a+95)^2 - 4^2$

$[\Theta\ ab = \left(\dfrac{a+b}{2}\right)^2 - \left(\dfrac{a-b}{2}\right)^2]$

$= \left(x^2 + 20x + 95\right)^2 - 4^2$ **(Ans.)**

PRACTICE SET 6:

Problem 1:

$a + b + c = 0$

$\Rightarrow a = -(b+c),\ b = -(c+a),\ c = -(a+b)$

$\therefore a^2 - bc = (-b-c)^2 - bc = b^2 + c(b+c) = b^2 - ca$

Similarly $b^2 - ca = (-c-a)^2 - ca = c^2 + a(c+a) = c^2 - ab$

$\therefore a^2 - bc = b^2 - ca = c^2 - ab$ **(Ans.)**

Problem 2:

$3\left(a^2 + b^2 + c^2\right) = (a+b+c)^2 = a^2 + b^2 + c^2 + 2(ab+bc+ca)$

$\Rightarrow 2\left(a^2 + b^2 + c^2\right) - 2(ab+bc+ca) = 0$

$$\Rightarrow (c-a)^2 + (b-c)^2 + (a-b)^2 = 0$$

$$[\Theta \; a^2 + b^2 + c^2 - ab - bc - ca = \frac{1}{2}[(b-c)^2 + (c-a)^2 + (a-b)^2]]$$

If the sum of square terms is zero, each must be equal to zero.

$$\Rightarrow c - a = 0, \; b - c = 0, \; a - b = 0$$

$$\Rightarrow c = a, \; b = c, \; a = b \Rightarrow a = b = c \; \textbf{(Ans.)}$$

Problem 3:

$$x^4 + \frac{1}{x^4} = 47$$

$$\Rightarrow \left(x^2 + \frac{1}{x^2}\right)^2 - 2.x^2.\frac{1}{x^2} = 47 \quad \Rightarrow \left(x^2 + \frac{1}{x^2}\right)^2 = 49 \quad \Rightarrow \left(x^2 + \frac{1}{x^2}\right) = \pm 7$$

$$\Rightarrow \left(x + \frac{1}{x}\right)^2 - 2.x.\frac{1}{x} = \pm 7$$

$$\Rightarrow \left(x + \frac{1}{x}\right)^2 = 9, \; \left(x + \frac{1}{x}\right)^2 = -5 \; \dots (1)$$

Again, $x^4 + \frac{1}{x^4} = 47$

$$\Rightarrow \left(x^2 - \frac{1}{x^2}\right)^2 + 2.x^2.\frac{1}{x^2} = 47 \quad \Rightarrow \left(x^2 - \frac{1}{x^2}\right)^2 = 45$$

$$\Rightarrow \left(x + \frac{1}{x}\right)^2 \left(x - \frac{1}{x}\right)^2 = 45$$

$$\Rightarrow \left(x - \frac{1}{x}\right)^2 = 5, \; \left(x - \frac{1}{x}\right)^2 = -9 \; \textbf{(Ans.)}$$

Problem 4:

$$\frac{1}{1-x} + \frac{1}{1-y} + \frac{1}{1-z}$$

$$\Rightarrow \frac{3 - 2(x+y+z) + (xy+yz+zx)}{(1-x)(1-y)(1-z)} = \frac{3 - 2.6 + 9}{(1-x)(1-y)(1-z)} = 0 \; \textbf{(Ans.)}$$

$$[\Theta \; x + y + z = 6, \; xy + yz + zx = 9]$$

Problem 5:

$$xy(x+y)=1 \Rightarrow (x+y)^3 = \frac{1}{x^3 y^3}$$

$$\Rightarrow x^3 + y^3 + 3xy\left(\frac{1}{xy}\right) = \frac{1}{x^3 y^3}$$

$$\Rightarrow \frac{1}{x^3 y^3} - x^3 - y^3 = 3 \text{ (Ans.)}$$

Problem 6:

$$x^2\left(\frac{x^2}{y}+y\right) + y^2\left(\frac{y^2}{x}+x\right)$$

$$= \left(x^2 + y^2\right)\left(\frac{x^2}{y}+\frac{y^2}{x}\right) = \left\{(x+y)^2 - 2xy\right\}\left\{\frac{x^3+y^3}{xy}\right\}$$

$$= \left\{(x+y)^2 - 2xy\right\}\left\{\frac{(x+y)^3 - 3xy(x+y)}{xy}\right\} = 655.2 \text{ (Ans.)}$$

$$[\Theta \,(x+y)=6\,, xy=5]$$

Problem 7:

Suppose $(x+2y)=b$, $(x+3y)=a$

$$\therefore (x+2y)(x+3y)= ab = \left(\frac{a+b}{2}\right)^2 - \left(\frac{a-b}{2}\right)^2 = \left\{\frac{1}{2}(2x+5y)\right\}^2 - \left(\frac{1}{2}y\right)^2 \text{ (Ans.)}$$

Problem 8:

$$\frac{(a+b)^2 + (b+c)^2 + (c+a)^2}{a^2 + b^2 + c^2}$$

$$= \frac{2\left(a^2 + b^2 + c^2 + bc + ca + ab\right)}{(a+b+c)^2 - 2(bc+ca+ab)} \quad [\Theta \,(a+b+c)^2 = a^2 + b^2 + c^2 + 2(bc+ca+ab)]$$

$$= \frac{2\left\{(a+b+c)^2 - (ab+bc+ca)\right\}}{(a+b+c)^2 - 2(bc+ca+ab)} = \frac{6}{2} = 3 \text{ (Ans.)}$$

$$[\Theta \, a+b+c = 2\,, ab+bc+ca = 1]$$

Problem 9:

$$a^4 - b^4 = (a^2)^2 - (b^2)^2 = (a^2 + b^2)(a^2 - b^2) = (a^2 + b^2)(a+b)(a-b)$$

$$a^2 + b^2 = \frac{1}{2}\{(a+b)^2 + (a-b)^2\} = 2 \quad [\Theta\, a - b = 1, a + b = \sqrt{3}]$$

$$\therefore a^4 - b^4 = 2.\sqrt{3}.1 = 2\sqrt{3} \textbf{ (Ans.)}$$

Problem 10:

$$u = x + \frac{1}{x} \Rightarrow u^2 - 2 = x^2 + \frac{1}{x^2}$$

$$\therefore x^4 + \frac{1}{x^4} = \left(x^2 + \frac{1}{x^2}\right)^2 - 2.x^2.\frac{1}{x^2} = (u^2 - 2)^2 - 2 = u^4 - 4u^2 + 2 \textbf{ (Ans.)}$$

PRACTICE SET 7:

Problem 1:

$$(a+b)^2 = a^2 + b^2 + 2ab = 58 + 42 = 100 \quad [\Theta\, a^2 + b^2 = 58, ab = 21]$$

$$\therefore a + b = \pm 10 \ \ldots.(1)$$

$$(a-b)^2 = a^2 + b^2 - 2ab = 58 - 42 = 16$$

$$\therefore a - b = \pm 4 \ \ldots.(2)$$

$$\therefore a = 7, b = 3;\ a = 3, b = 7;\ a = -7, b = -3;\ a = -3, b = -7 \textbf{ (Ans.)}$$

Problem 2:

$$xy + yz + zx = \frac{1}{2}(2xy + 2yz + 2zx)$$

$$= \frac{1}{2}\{(x + y + z)^2 - (x^2 + y^2 + z^2)\} = \frac{1}{2}(a^2 - b) \textbf{ (Ans.)}$$

$$[\Theta\, x + y + z = a,\ x^2 + y^2 + z^2 = b]$$

Problem 3:

$$\Theta\, a + b + c = 8 \text{ and } a^2 + b^2 + c^2 = 30$$

$$\therefore ab + bc + ca = \frac{1}{2}(2ab + 2bc + 2ca) = \frac{1}{2}\{(a + b + c)^2 - (a^2 + b^2 + c^2)\} = \frac{1}{2}(64 - 30) = 17$$

$$\therefore a^3 + b^3 + c^3 - 3abc = (a+b+c)(a^2 + b^2 + c^2 - ab - bc - ca)$$
$$= 8(30 - 17) = 104 \text{ (Ans.)}$$

Problem 4:

$$a = 1 - \frac{1}{b} \Rightarrow a = \frac{b-1}{b}$$

$$b = \frac{2}{c} - 1 \Rightarrow c = \frac{2}{b+1}$$

$$\therefore a + b + c$$

$$= \frac{b-1}{b} + b + \frac{2}{b+1}$$

$$= \frac{(b-1)(b+1) + b^2(b+1) + 2b}{b(b+1)}$$

$$= \frac{b^2 - 1 + b^3 + b^2 + 2b}{b(b+1)}$$

$$= \frac{b^3 + 2b^2 + 2b - 1}{b(b+1)} = 2 + \frac{b^3 - 1}{b(b+1)} \text{ (Ans.)}$$

Problem 5:

Assume, $27x^3 - 9ax^2 + 36x + b = (3x - m)^3$

$$\Rightarrow 27x^3 - 9ax^2 + 36x + b = 27x^3 - 27x^2 m + 9xm^2 - m^3$$

Since this is an identity, the corresponding terms must be equal.

$$\therefore -9ax^2 = -27x^2 m \quad \dots (1)$$
$$36x = 9xm^2 \quad \dots (2)$$
$$b = -m^3 \quad \dots (3)$$

From (2), $9m^2 = 36 \Rightarrow m = \pm 2$

Case1: $m = +2$

  From (3), $b = -8$

  From (1), $a = 6$

Case2: $m = -2$

  From (3), $b = 8$

  From (1), $a = -6$

The values of $a = 6, -6$ and $b = -8, 8$ **(Ans.)**

Problem 6:

$$a^4 + 2a^3b - 2ab^3 - b^4$$
$$= (a^2 + b^2)(a^2 - b^2) + 2ab(a^2 - b^2)$$
$$= (a^2 - b^2)(a^2 + b^2 + 2ab)$$
$$= (a-b)(a+b)(a+b)(a+b) \quad \textbf{(Ans.)}$$

Problem 7:

$$x^3 + 15x^2 + 75x + 225$$
$$= (x^3 + 3.x^2.5 + 3x.5^2 + 5^3) - 5^3 + 225$$
$$= (x+5)^3 - 125 + 225 = (x+5)^3 + 100 = (\sqrt[3]{5} - 5 + 5)^3 + 100 = 105$$
$$[\Theta \, x = \sqrt[3]{5} - 5]$$
$$x^3 + 15x^2 + 75x + 119 = (x^3 + 15x^2 + 75x + 125) - 6$$
$$= (x+5)^3 - 6 = (\sqrt[3]{5})^3 - 6 = 5 - 6 = -1 \quad [\Theta \, x = \sqrt[3]{5} - 5]$$
$$\therefore \frac{x^3 + 15x^2 + 75x + 225}{x^3 + 15x^2 + 75x + 119} = -\frac{105}{1} = -105 \quad \textbf{(Ans.)}$$

Problem 8:

$$x = b - c, \; y = c - a, \; z = a - b$$
$$\Rightarrow x + y + z = 0$$
$$\therefore x^2 - y^2 + z^2 + 2zx = (x+z)^2 - y^2 = (x+y+z)(x-y+z) = 0 \quad \textbf{(Ans.)}$$

Problem 9:

$$x^3 + y^3 = (x+y)^3 - 3xy(x+y) = (1+xy)^3 - 3xy(1+xy) \quad [\Theta \, x + y = 1 + xy]$$
$$= 1 + x^3y^3 + 3xy(1+xy) - 3xy(1+xy)$$
$$= 1 + x^3y^3 \quad \textbf{(Ans.)}$$

Problem 10:

$$\frac{a}{b} + \frac{b}{a} + \frac{c}{b} + \frac{b}{c} + \frac{a}{c} + \frac{c}{a} + 3$$
$$= \left(\frac{a}{b} + \frac{c}{b} + 1\right) + \left(\frac{b}{a} + \frac{c}{a} + 1\right) + \left(\frac{a}{c} + \frac{b}{c} + 1\right)$$

$$= \left(\frac{a+b+c}{b}\right) + \left(\frac{a+b+c}{b}\right) + \left(\frac{a+b+c}{b}\right)$$

$$= (a+b+c)\left(\frac{1}{a}+\frac{1}{b}+\frac{1}{c}\right) \textbf{ (Ans.)}$$

PRACTICE SET 8:

Problem 1:

$$x + \frac{4}{x} = 4$$

$$\Rightarrow x^2 - 4x + 4 = 0$$

$$\Rightarrow (x-2)^2 = 0$$

$$\Rightarrow x = 2$$

$$\therefore x^3 + \frac{4}{x^3} = 8\frac{1}{2} \textbf{ (Ans.)}$$

Problem 2:

$$x^4 + \frac{1}{x^4} = \left(x^2 - \frac{1}{x^2}\right)^2 + 2.x^2.\frac{1}{x^2} = \left(x + \frac{1}{x}\right)^2 \left(x - \frac{1}{x}\right)^2 + 2$$

$$= \left\{\left(x - \frac{1}{x}\right)^2 + 4\right\}\left(x - \frac{1}{x}\right)^2 + 2 = 119 \textbf{ (Ans.)}$$

Problem 3:

$$\frac{1}{1-x} + \frac{1}{1+x} + \frac{2}{1+x^2} + \frac{4}{1+x^4} - \frac{8}{1-x^8}$$

$$= \frac{2}{1-x^2} + \frac{2}{1+x^2} + \frac{4}{1+x^4} - \frac{8}{1-x^8}$$

$$= \frac{4}{1-x^4} + \frac{4}{1+x^4} - \frac{8}{1-x^8}$$

$$= \frac{8}{1-x^8} - \frac{8}{1-x^8} = 0 \textbf{ (Ans.)}$$

Problem 4:

$$(a^2 - b^2)(x^2 + y^2) + 2(a^2 + b^2)xy$$
$$= a^2x^2 + a^2y^2 - b^2x^2 - b^2y^2 + 2a^2xy + 2b^2xy$$
$$= (a^2x^2 + a^2y^2 + 2a^2xy) - (b^2x^2 + b^2y^2 - 2b^2xy)$$
$$= (ax + ay)^2 - (bx - by)^2$$
$$= (ax + ay + bx - by)(ax + ay - bx + by) \textbf{ (Ans.)}$$

Problem 5:

$$a^3 + \frac{1}{a^3} = \left(a + \frac{1}{a}\right)^3 - 3.a.\frac{1}{a}\left(a + \frac{1}{a}\right) = \left(a + \frac{1}{a}\right)\left(a + \frac{1}{a}\right)^2 - 3\left(a + \frac{1}{a}\right)$$

$$= 3\left(a + \frac{1}{a}\right) - 3\left(a + \frac{1}{a}\right) = 0 \textbf{ (Ans.)}$$

$$[\Theta\left(a + \frac{1}{a}\right)^2 = 3]$$

Problem 6:

Assume $2x - 1 = a$, $x - 2 = b$

$$\Rightarrow a - b = x + 1$$

$$\therefore (2x - 1)^3 - (x - 2)^2 - 3(2x - 1)(x - 2)(x + 1)$$

$$= a^3 - b^3 - 3ab(a - b) = (a - b)^3 = (x + 1)^3 \textbf{ (Ans.)}$$

Problem 7:

$$\left(x - \frac{1}{x}\right)^2 = 3$$

$$\therefore \left(x + \frac{1}{x}\right)^2 = \left(x - \frac{1}{x}\right)^2 + 4 = 7$$

$$\therefore x^6 + \frac{1}{x^6} = \left(x^3\right)^2 + \left(\frac{1}{x^3}\right)^2$$

$$= \left(x^3 + \frac{1}{x^3}\right)^2 - 2 \quad [\Theta\, a^2 + b^2 = (a + b)^2 - 2ab]$$

$$= \left[\left(x + \frac{1}{x} \right)^3 - 3\left(x + \frac{1}{x} \right) \right]^2 - 2 \quad [\Theta\ a^3 + b^3 = (a+b)^3 - 3ab(a+b)]$$

$$= \left(x + \frac{1}{x} \right)^2 \left[\left(x + \frac{1}{x} \right)^2 - 3 \right]^2 - 2$$

$$= 7 \times (4)^2 - 2 = 110 \ \textbf{(Ans.)}$$

Problem 8:

$$\left(a^2 - b^2 \right)^2 = (a+b)^2(a-b)^2 = 1.\left(a^2 - 2ab + b^2 \right) = 1.\left(a^2 - ab + b^2 \right) - ab \quad [\Theta\ a+b=1]$$
$$= (a+b)\left(a^2 - ab + b^2 \right) - ab = a^3 + b^3 - ab \ \textbf{(Ans.)}$$

Problem 9:

$$(a+b+c)^2 = a^2 + b^2 + c^2 + 2(bc + ca + ab) = 25$$
$$\Rightarrow (a+b+c) = \pm 5 \ \textbf{(Ans.)}$$

Problem 10:

$$\frac{x^3 - y^3 - 12xyz}{(x-y)^2} = \frac{x^3 - y^3 - 3xy.4z}{(x-y)^2} = \frac{x^3 - y^3 - 3xy(x-y)}{(x-y)^2} = (x-y) = 4z \ \textbf{(Ans.)}$$

PRACTICE SET 9:

Problem 1:

$$x + y + z = b + c - a + c + a - b + a + b - c = a + b + c$$
$$x - y = b + c - a - c - a + b = 2(b-a)$$
$$y - z = c + a - b - a - b + c = 2(c-b)$$
$$z - x = a + b - c - b - c + a = 2(a-c)$$

$$\therefore x^3 + y^3 + z^3 - 3xyz = (x+y+z)\left(x^2 + y^2 + z^2 - xy - yz - zx \right)$$
$$= (x+y+z).\frac{1}{2}\left\{ 2x^2 + 2y^2 + 2z^2 - 2xy - 2yz - 2zx \right\}$$

$$= \frac{1}{2}(x+y+z)\{(x-y)^2 + (y-z)^2 + (z-x)^2\}$$

$$= \frac{1}{2}(a+b+c)\{4(b-a)^2 + 4(c-b)^2 + 4(a-c)^2\}$$

$$= 4(a+b+c)\{a^2 + b^2 + c^2 - bc - ca - ab\}$$

$$= 4(a^3 + b^3 + c^3 - 3abc) \text{ (Ans.)}$$

Problem 2:

$$(a-1)x^2 + a^2xy + (a+1)y^2$$

$$= (a-1)x^2 + (a^2-1)xy + xy + (a+1)y^2$$

$$= (a-1)x\{x+(a+1)y\} + y\{x+(a+1)y\}$$

$$= \{x+(a+1)y\}\{(a-1)x+y\}$$

$$= (x+ay+y)(ax-x+y) \text{ (Ans.)}$$

Problem 3:

$$(x+4)(x^2-4x+16)-(x+3)(x^2-3x+9)$$

$$= (x+4)(x^2-4.x+4^2)-(x+3)(x^2-3.x+3^2)$$

$$= (x^3+4^3)-(x^3+3^3) = 37 \text{ (Ans.)}$$

$$[\Theta (a+b)(a^2-ab+b^2)=a^3+b^3]$$

Problem 4:

$$x^3 - y^3 + 3y^2 - 3y + 1$$

$$= x^3 - (y^3 - 3y^2 + 3y - 1)$$

$$= (x)^3 - (y-1)^3$$

$$= (x-y+1)\{x^2 + x(y-1) + (y-1)^2\}$$

$$= (x-y+1)(x^2 + y^2 + 1 + xy - x - 2y) \text{ (Ans.)}$$

Problem 5:

$$\left(x-\frac{1}{x}\right)^2 = 3 \implies x - \frac{1}{x} = \sqrt{3}$$

$$\therefore x + \frac{1}{x} = \sqrt{\left(x - \frac{1}{x}\right)^2 + 4.x.\frac{1}{x}} = \sqrt{7}$$

$$\therefore x^3 + \frac{1}{x^3} = \left(x + \frac{1}{x}\right)^3 - 3.x.\frac{1}{x}\left(x + \frac{1}{x}\right) = 4\sqrt{7} \quad \textbf{(Ans.)}$$

Problem 6:

$$\left(a^2 - b^2\right)\left(x^2 - y^2\right) + 4abxy$$

$$= a^2 x^2 - a^2 y^2 - b^2 x^2 + b^2 y^2 + 4abxy$$

$$= \left(a^2 x^2 + 2abxy + b^2 y^2\right) - \left(a^2 y^2 - 2abxy + b^2 x^2\right)$$

$$= (ax + by)^2 - (ay - bx)^2$$

$$= (ax + by + ay - bx)(ax + by - ay + bx) \quad \textbf{(Ans.)}$$

Problem 7:

$$\left(a^{1/3} + b^{1/3} + c^{1/3}\right) = 0$$

$$\therefore \left(a^{1/3}\right)^3 + \left(b^{1/3}\right)^3 + \left(c^{1/3}\right)^3 = 3.a^{\frac{1}{3}} b^{\frac{1}{3}} . c^{\frac{1}{3}}$$

$$\Rightarrow a + b + c = 3.a^{\frac{1}{3}} b^{\frac{1}{3}} . c^{\frac{1}{3}}$$

$$\Rightarrow (a + b + c)^3 = 27abc$$

$$\Rightarrow (a + b + c) = 3 \quad \textbf{(Ans.)}$$

Problem 8:

$$(y + z)(y - z)\left(y^2 + z^2\right) + (z + x)(z - x)\left(z^2 + x^2\right) + (x + y)(x - y)\left(x^2 + y^2\right)$$

$$= \left(y^2 - z^2\right)\left(y^2 + z^2\right) + \left(z^2 - x^2\right)\left(z^2 + x^2\right) + \left(x^2 - y^2\right)\left(x^2 + y^2\right)$$

$$= y^4 - z^4 + z^4 - x^4 + x^4 - y^4 = 0 \quad \textbf{(Ans.)}$$

Problem 9:

$$x^3 + y^3 + 297xy$$

$$= x^3 + y^3 + 3.xy.99 = x^3 + y^3 + 3.xy.(x + y) = (x + y)^3 = 99^3 = 970299 \quad \textbf{(Ans.)}$$

Problem 10:

$$x(x-3)(x-6)(x-9)+81$$
$$= x(x-9)(x-3)(x-6)+81$$
$$= (x^2 -9x)(x^2 -9x+18)+81$$
$$= a(a+18)+81 = a^2 +18a+81 = (a+9)^2 = (x^2 -9x+9)^2 \ \textbf{(Ans.)}$$

PRACTICE SET 10:

Problem 1:

$$(x+1)(x+3)(x+5)(x+7)+15$$
$$= \{(x+1)(x+7)\}\{(x+3)(x+5)\}+15 \ [\text{Rearranging terms}]$$
$$= (x^2 +8x+7)(x^2 +8x+15)+15$$
$$= (a+7)(a+15)+15 \ [\text{Assume } a = x^2 +8x]$$
$$= a^2 +22a+120$$
$$= a^2 +2.a.11+121-1$$
$$= (a+11)^2 -1^2$$
$$= (a+12)(a+10)$$
$$= (x^2 +8x+12)(x^2 +8x+10) = (x+2)(x+6)(x^2 +8x+10) \ \textbf{(Ans.)}$$

Problem 2:

$$(x+y+z)(x-y+z)(x+y-z)(z+y-x)$$
$$= \{(x+y)^2 -z^2\}\{z^2 -(x-y)^2\}$$
$$= (x^2 +y^2 -z^2 +2xy)(z^2 -x^2 -y^2 +2xy)$$
$$= 2x^2 y^2 +2y^2 z^2 +2z^2 x^2 -x^4 -y^4 -z^4 \ \textbf{(Ans.)}$$

Problem 3:

Assume $a-b=x$, $b-c=y$, $c-a=z$

$$\therefore x+y+z=0$$

$$\frac{1}{16a^2} + \frac{(a-b)^2 +(b-c)^2 +(c-a)^2}{(a-b)(b-c)(c-a)} + \frac{2}{a-b} + \frac{2}{b-c} + \frac{2}{c-a} +49a^2$$

$$= 49a^2 + \frac{1}{16a^2} + \frac{x^2 + y^2 + z^2}{xyz} + \frac{2}{x} + \frac{2}{y} + \frac{2}{z}$$

$$= (7a)^2 + \left(\frac{1}{4a}\right)^2 + \frac{x^2 + y^2 + z^2 + 2yz + 2xz + 2xy}{xyz}$$

$$= \left(7a + \frac{1}{4a}\right)^2 - 2.7a.\frac{1}{4a} + \frac{(x+y+z)^2}{xyz} = \frac{1}{2} + 0 = \frac{1}{2} \text{ (Ans.)}$$

$$\left[\Theta\ 7a + \frac{1}{4a} = -2 \ \text{ and } \ (x+y+z)^2 = x^2 + y^2 + z^2 + 2yz + 2xz + 2xy\right]$$

Problem 4:

$$x + y = 2a \ \& \ y = a - \frac{1}{a}$$

$$\Rightarrow x - y = \frac{2}{a}$$

$$\therefore \frac{x^4 + y^4 - 2x^2 y^2}{(x+y)^3} = \frac{(x^2 - y^2)^2}{(x+y)^3} = \frac{(x+y)^2 (x-y)^2}{(x+y)^3} = \frac{(x-y)^2}{(x+y)} = \frac{4}{a^2 .2a} = \frac{2}{a^3} \text{ (Ans.)}$$

Problem 5:

$$a^2 + b^2 = (a+b)^2 - 2ab = 5$$

$$a^6 + b^6 = (a^2)^3 + (b^2)^3 = (a^2 + b^2)^3 - 3a^2 b^2 (a^2 + b^2) = 5^3 - 3.4.5 = 65 \text{ (Ans.)}$$

$$[\Theta\ a + b = 3 \ \& \ ab = 2]$$

Problem 6:

$$(x+1)(x+3)(x-5)(x-7)$$

$$= (x+1)(x-5)(x+3)(x-7)$$

$$= (x^2 - 4x - 5)(x^2 - 4x - 21)$$

$$= (a - 5)(a - 21)$$

$$= a^2 - 26a + 105 = (a - 13)^2 - 8^2 = (x^2 - 4x - 13)^2 - 8^2 \text{ (Ans.)}$$

Problem 7:

$$x = \sqrt{3} - \frac{1}{\sqrt{3}} \ \ldots.(1) \qquad y = \sqrt{3} + \frac{1}{\sqrt{3}} \ \ldots.(2)$$

Adding (1) and (2), $x + y = 2\sqrt{3}$.

Multiplying (1) and (2), $xy = \dfrac{8}{3}$.

$$\therefore \frac{x^2}{y} + \frac{y^2}{x} = \frac{x^3 + y^3}{xy} = \frac{(x+y)^3 - 3xy(x+y)}{xy} = 3\sqrt{3} \textbf{ (Ans.)}$$

Problem 8:

$$\left(x^2 - 1\right)(x+2)x - 8$$
$$= (x+1)(x-1)(x+2)x - 8$$
$$= x(x+1)(x-1)(x+2) - 8$$
$$= \left(x^2 + x\right)\left(x^2 + x - 2\right) - 8$$
$$= a(a-2) - 8 = a^2 - 2a - 8 = \left(a^2 - 2a + 1\right) - 3^2 = (a-1)^2 - 3^2$$
$$= (a+2)(a-4) = \left(x^2 + x + 2\right)\left(x^2 + x - 4\right) \textbf{ (Ans.)}$$

Problem 9:

$$\left(x - \frac{1}{x}\right)^2 = 3 \Rightarrow \left(x - \frac{1}{x}\right) = \sqrt{3}$$

$$\therefore x^3 - \frac{1}{x^3} = \left(x - \frac{1}{x}\right)^3 + 3.x.\frac{1}{x}\left(x - \frac{1}{x}\right) = \left(\sqrt{3}\right)^3 + 3.1.\sqrt{3} = 6\sqrt{3}$$

$$\therefore x^6 + \frac{1}{x^6} = \left(x^3 - \frac{1}{x^3}\right)^2 + 2.x^3.\frac{1}{x^3} = \left(6\sqrt{3}\right)^2 + 2 = 108 + 2 = 110 \textbf{ (Ans.)}$$

Problem 10:

$$(x+1)(x+2)(3x-1)(3x-4) + 12$$
$$= \{(x+1)(3x-1)\}\{(x+2)(3x-4)\} + 12$$
$$= \left(3x^2 + 2x - 1\right)\left(3x^2 + 2x - 8\right) + 12$$
$$= (a-1)(a-8) + 12 \ [\text{Assume } 3x^2 + 2x = a]$$
$$= a^2 - 9a + 8 + 12$$
$$= a^2 - 9a + 20 = (a-5)(a-4)$$
$$= \left(3x^2 + 2x - 5\right)\left(3x^2 + 2x - 4\right)$$
$$= (x-1)(3x+5)\left(3x^2 + 2x - 4\right) \textbf{ (Ans.)}$$

ABOUT THE AUTHOR

Sanjay Jamindar, a Wireless Telecommunication Engineer and Consultant by profession, has to his credit many a technical contribution to various clientele in the field of Telecommunication within the Information Technology industry. During his professional career, he published many technical papers in national and international conferences.

Sanjay has a Bachelor's degree in Physics (Hons.) and Bachelor's in Engineering (Electronics & Communication) from Calcutta University, Kolkata, India. He completed his Master's (M.Tech) in Laser Technology and Optical Communication from Indian Institute Of Technology (IIT), Kanpur, India.

He received National Scholarships during Schooling and Bachelor's degree (Physics) Examinations. He also received DAAD (German Govt.) Scholarship for pursuing Master's thesis work in Germany (Technical University, Berlin) under DAAD-IIT Exchange Programme.

Sanjay's interest and passion lies in research in the Telecommunication domain. He also writes technical books including various educational books that can help students in strengthening their knowledge, in the domain of Mathematics and Physics.

www.ingramcontent.com/pod-product-compliance
Lightning Source LLC
Chambersburg PA
CBHW081314170526
45166CB00011B/3524